Mutlaq Alotaibi

Security Aspects for VoIP Systems

AF153315

Mutlaq Alotaibi

Security Aspects for VoIP Systems

LAP LAMBERT Academic Publishing

Imprint

Any brand names and product names mentioned in this book are subject to trademark, brand or patent protection and are trademarks or registered trademarks of their respective holders. The use of brand names, product names, common names, trade names, product descriptions etc. even without a particular marking in this work is in no way to be construed to mean that such names may be regarded as unrestricted in respect of trademark and brand protection legislation and could thus be used by anyone.

Cover image: www.ingimage.com

Publisher:
LAP LAMBERT Academic Publishing
is a trademark of
Dodo Books Indian Ocean Ltd. and OmniScriptum S.R.L publishing group

120 High Road, East Finchley, London, N2 9ED, United Kingdom
Str. Armeneasca 28/1, office 1, Chisinau MD-2012, Republic of Moldova, Europe
Managing Directors: Ieva Konstantinova, Victoria Ursu
info@omniscriptum.com

Printed at: see last page
ISBN: 978-3-659-71678-2

Zugl. / Approved by: Leicester,De montfort university,2014

Abstract

VoIP is now considered one of the most commonly deployed technologies all over the world. Its applications are now widely used by different people from all ages. Ensuring security within these applications is considered an essential factor in making it more reliable and encouraging people to increase adopting VoIP applications. During this project; a secure peer to peer voice call communication system will be developed. This will be achieved by employing three encryption methods; DES, 3-DES and AES algorithms. A JAVA program using these three methods will be developed for the encryption and decryption stages. This program is run on the client sides in order to establish secure VoIP connection for the voice data. This means that higher level of security can be achieved by encrypting voice data itself; so it cannot be recognized unless applying decryption process. A comparison is also carried out between the three applied encryption methods based on the encryption and decryption delays. It was noted that the time taken by encryption and decryption per packet does not exceed 1ms for the three methods when run on a laptop with Corei7 CPU (3GHz) and 4GB RAM.

Acknowledgements

I would like to take this opportunity to thank everyone who helped and supported me during my work. First of all, I thank the Almighty ALLAH who helped me reach this stage.

I would also like to thank my first supervisor Dr. Shakeel Ahmad for his support, help and guidance which he kindly provided during the course of my research. Prof. Raouf Hamzaoui's feedback on my work also helped me to improve it. For this, I am very thankful to him.

My thanks are also to my family and friends who kept on encouraging and supporting me during this thesis. I would especially like to mention the names of my father Bader Al Otaibi, my mother Fatimah Al Otaibi, my wife Samirah, my son Bader, and my daughters Fatoun, Ghadnaa, Ghaida, and Jumanah,

Lastly, I would like to thank Mr. Abdullah Al Majid who encouraged me to pursue higher studies and supported me in every way.

A big thank you is also due to all my friends and colleagues with whom I studied and relaxed and with whom I shared some unforgettable moments of my life.

Table of Contents

List of Figures

List of Tables

List of Publications

M. Alotaibi, S. Ahmad, A. AlMajid, and R. Hamzaoui, "Security Aspects in Voice over Internet Protocol (VoIP) Systems: Review Paper " (in preparation).

List of Abbreviations

AES	Advanced Encryption Standard
DES	Data Encryption Standard
3-DES	Triple Data Encryption Standard
DHCP	Dynamic Host Configuration Protocol
DKE	Dynamic Key Exchange
HTTTP	Hyper Text Transfer Protocol
IMS	Internet Multimedia Subsystem
IAX	Inter-Asterisk Exchange
IEFT	Internet Engineering Force Task
ITU	International Telecommunication Union
IVR	Interactive Voice Response
KEP	Key Exchange Protocol
MGCP	Media Gateway Control Protocol
MOS	Mean Opinion Score
NAT	Network Address Translation
RC4	Ron's Code 4
RSA	Rivest, Shamir, and Adelman
RTP	Real-time Transport Protocol
RFC	Request for Comments
RBAC	Role Based Access Control
SCCP	Skinny Client Control Protocol
SDP	Session Description Protocol
SIP	Session Initiation Protocol
SMTP	Simple Mail Transfer Protocol

SPIT	Spam over IP Telephony
STCP	Stream Control Transmission Protocol
PBX	Private Branch Exchange
PSTN	Public Switched Telephone Network
QoS	Quality of Service
TCP	Transmission Control Protocol
TLS	Transport Layer Security
UDP	User Datagram Protocol
VoIP	Voice over Internet Protocol
XMPP	Extensible Messaging and Presence Protocol

Chapter One: Introduction

1.1   Overview

Voice-Over Internet Protocol (VoIP) is the current standard for voice communication via the Internet. This technology uses available IP networks and minimises communication costs by utilising the convenient Public Switched Telephone Network (PSTN). Many factors encourage people to adopt VoIP as an effective solution for voice or video communication purposes in the Internet environment, such as the reduction of hardware necessary for communication and ease of deployment. In addition, VoIP provides personalised services and value-added flexibility, which enables customised solutions for voice and video exchange. [1].

VoIP services are used for making voice calls through internet by using IP. This technology is characterised by several advantages—such as no fees—that indicate that it will be the future for voice calls in the next generation of communications. The voice calls are performed in VoIP systems via transferring the voice in form of digital packets through internet. This will in turns enable novel practical VoIP applications, such as; Skype, Viber and yahoo messenger [2].

Today, VoIP technology is rapidly and extensively deployed. Efficiency, low costs and flexibility are the main factors that make VoIP attractive to companies. However, problems of security may develop from extensive deployment of VoIP. Therefore, this project reviews the systems and security issues of VoIP, specifically in the context of it being used for the media transport and signalling between clients. Mechanisms of defence against security vulnerabilities and threats are described, along with current and potential attacks that face VoIP systems and the approaches being adopted to address these attacks [1].

1.2   VoIP Protocols

Several protocols can be used to manage and control the data exchange process between a transmitter and a receiver in a VoIP system. The most commonly used protocols will be introduced and briefly discussed within this section.

1.2.1   Session Initiation Protocol

The Session Initiation Protocol (SIP) is the most common protocol that is deployed in VoIP systems. This protocol has been made uniform through the Internet Engineering Force Task (IEFT). SIP/2.0 is the current version of this protocol; the codec for this version is performed via a request for comments 3261. Both Simple Mail Transfer Protocol (SMTP) and Hyper Text Transfer Protocol (HTTP) are included within SIP. SIP is an example of a signalling type of protocol; this means that the voice data transmission is not actually performed by SIP. The idea behind this signalling protocol is initiating, organising and then terminating the communication session between the transmitter and the receiver. SIP is also responsible for phone rings and the busy tone in addition to ending calls [3] [4].

1.2.2   Real-time Transport Protocol

The Real-Time Transport Protocol (RTP) is also a commonly used protocol in the VoIP environment, especially for conveying real-time data. This protocol is used for voice transmission in addition to the transmission of any data type, such as video, audio, etc. RTP can also be used for multimedia streaming. Despite the fact that there are several protocols that can be used in place of SIP, both SIP and RTP are usually deployed together to perform the whole communication process within VoIP, as illustrated in Figure 1 [3].

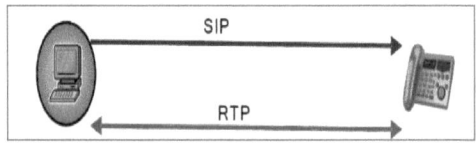

Figure 1: VoIP call deploying both RTP and SIP protocols [3].

1.2.3   Other Signaling Protocols

Several other protocols can be used as alternatives for SIP. The first one is an H.323 protocol that has been standardised by the International Telecommunication Union. Despite the fact that this protocol can provide a user with approximately the same functionality that can be achieved with SIP, the deployment of SIP is still preferable since H.323 is old style. The second alternative is the Inter-Asterisk Exchange (IAX) protocol. This protocol was developed by the Asterisk Project Company and was initially employed by the Asterisk-type servers, which are VoIP servers. Both voice and signalling data are combined into only one stream of data that can be transferred using the User Datagram Protocol (UDP) through port number 4569. Another available alternative is the Skinny Client Control Protocol (SCCP), which is a proprietary protocol; it is provided by Cisco and deployed within servers for call management. It is also provided by the related terminals of phones. Jabber, which is an Extensible Messaging and Presence Protocol (XMPP) is also an alternative. This protocol has been made uniform through IEFT and is also used for instant messaging [3].

1.3   VoIP Architecture

The standard architecture for the VoIP system is shown in Figure 2 [5].

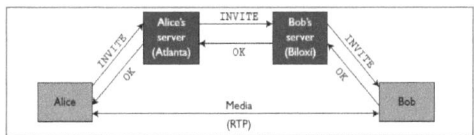

Figure 2: STANDARD ARCHITECTURE FOR VOIP SYSTEM [5].

As can be seen in Figure 2, a connection needs to be established between Alice and Bob. After Alice establishes this connection, the process of transmitting and receiving the call can be initiated. One of the most critical problems that must be addressed within VoIP systems is mapping the address of Bob, whether it is an email, phone number, etc., with the real machine that the user is deploying. This address can also be changed continuously. Alice can access Bob via either a desktop soft phone or a mobile phone [5].

This can be provided by deploying a rendezvous server network. Accounts are assigned for both Bob and Alice within permanent servers that can be accessed by everyone. In regard to the signalling process, if Bob and Alice want to contact each other, the first step of the rendezvous is deploying their own servers. The flow of the message is demonstrated in Figure 2. If Alice is planning ahead of time to communicate with Bob, then she should first contact the Bob's server and then request that an SIP INVITE request be sent to Bob. The most significant information that must be included within the INVITE request are the request target address, the ID of the call and the offer of the

Session Description Protocol (SDP), which contains the media session parameters established by Alice. After this is completed, then the media exchange process can be easily initiated [5].

1.4   Brief History of Security in VoIP

Security within VoIP has been studied by other researchers, who have also proposed designs and models for increasing the security and privacy of transferred data. The main issues that must be considered during the implementation of a VoIP system are introduced in [6]. The principles that should be employed to achieve security in a VoIP system are presented in [8]. In [13], a denial of service (DOS) attack in a VoIP system is addressed, and the main vulnerabilities of a VoIP system are discussed and introduced in [17].

The applications of secure protocols that can also achieve an acceptable level of security are introduced in [18] for secure RTP, and in [20], the weak points of using VoIP in general are discussed. Data encryption is one of the most common ways to achieve security in any communication system; this prevents a user's data from being recognised if it is accessed by unauthenticated users. [23] Discusses encryption options, such as Advanced Encryption Standard (AES), Data Encryption Standard (DES) and Ron's Code 4 (RC4), in addition to introducing a secure VoIP model-based encryption. These studies will be explained in more detail in the next chapter.

1.5   Problem Statement

VoIP services are now widely deployed all over the world. Several applications are available for users in different places and positions, such as organizations, companies and individual use. The main issue that limits user satisfaction regarding the adoption of this service is the lack of security. The VoIP technology is preferable to ordinary ones, such as PSTN because of its many advantages, such as very low costs, flexibility, integrity, better quality and a common structure for both Internet and voice services. However, since this technology is based on Internet use, it is vulnerable to the various threats that occur within the Internet environment, such as DOS attacks, spoofing and sniffing.

It is important to mention that security in VoIP systems can be provided on several levels, along media, channel protocols, or it can be provided by applying security mechanisms from the client's side, which will provide communication channels with secured data. It is impossible to know if data are being attacked unless a decryption key is applied, which results in an increased level of security. From the security viewpoint, delay is the main constraint, and it is preferable to minimise delays in real-time communications. Several issues pose the current nature of the problem. For the transmission of voice; the maximum allowable value for the delay within packets equals to 150 ms. This value can ensure the clarity of voice and it may be increased to 200 ms when applying encryption methods.

1.6   Scope and Objectives

This research proposes a solution for the problem of voice networks attacks by intruders by designing an enhanced VoIP system that can secure networks, prevent any illegal entrance and detect attacks. The purpose of this project is to integrate security features using multiple encryption methods, such as AES, DES or Triple DES (3DES) on the client's side when communicating through VoIP, and to measure the performance of a VoIP client that has integrated security or encryption modules. Developing a VoIP client that is integrated with the encryption module and still has good performance is the goal of this project. An integrated encryption module is expected to secure VoIP user communication. The steps taken to carry out this research are as follows:

1. Develop a VoIP communication system using open source software.
2. Develop modules to encrypt/decrypt voice data in real time.
3. Integrate encryption/decryption modules with the VoIP data.
4. Run the VoIP communication system with encryption/decryption enabled in real time.
5. Measure the effects of encryption/decryption on performance.

The following objectives must be met to achieve the main purpose of this project:

1. Study the structure, techniques, services, security protocols, services, security and encryption methods of VoIP.
2. Determine the protocols that are used in VoIP applications in addition to the methods of enumeration, methods and tools of attack, versions, vulnerabilities and deployment for VoIP services.
3. Enhance the security of the RTP protocols by adding an encryption module.
4. Apply the encryption methods to the system to improve performance and to

eliminate vulnerabilities.

5. Analyse the VoIP clients' performance in terms of end-to-end delay (the delay from when the sender speaks to when the receiver hears his voice) and in terms of the encryption/decryption delay.

1.7   Project Methodology

Figure 3 summarizes the project methodology.

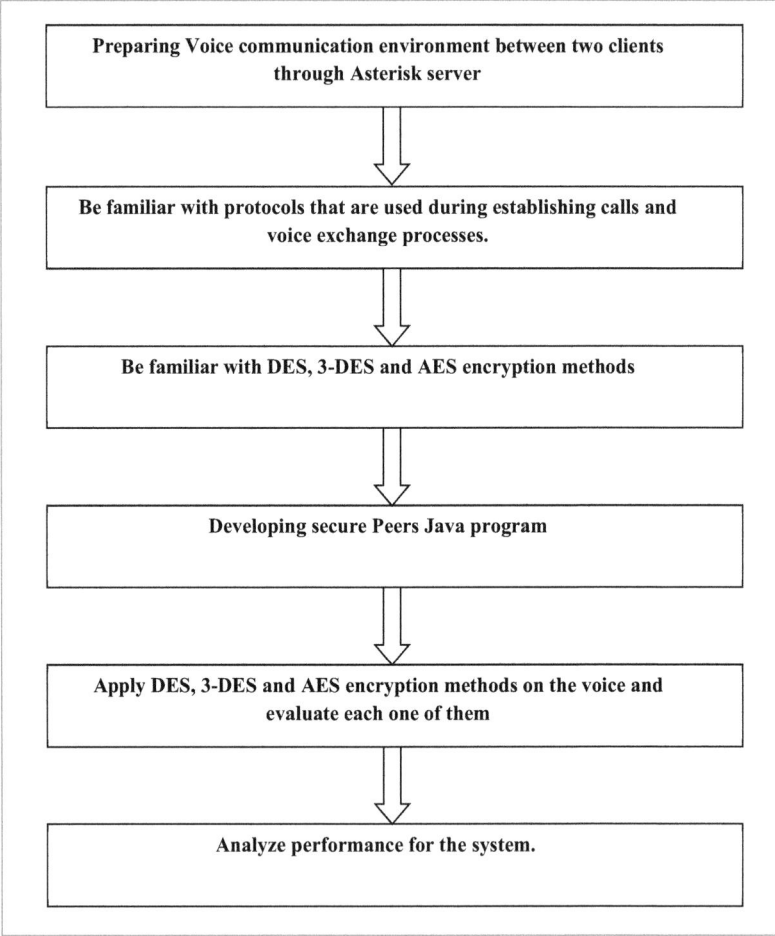

Figure 3: Project methodology.

As illustrated in Figure 3, the first step in this project is to establish the environment that will be used during the voice communication. This environment is illustrated in Figure 4.

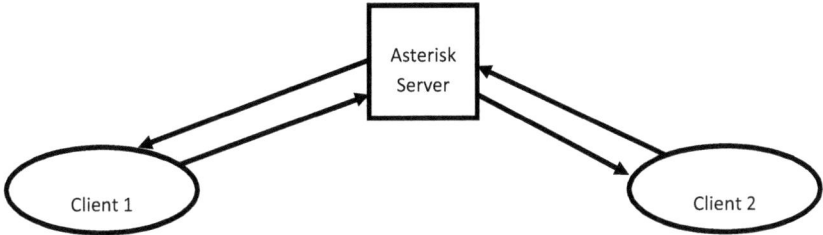

Figure 4:commuinication environment.

After establishing this environment, the general concepts for the voice communication process through the Asterisk server will be recognised including VoIP structure, call initiating, establishing, data exchange and call terminating. Once Client 1 wants to call Client 2, he must send a request to the server, and the server forwards it to Client 2. If Client 2 accepts this request, he will send an acknowledgment message to the server, and the server will forward it to Client 1. During this session, the SIP protocol is working. After establishing a call channel between the two clients, it then starts exchanging voice. The RTP protocol is working during this stage.

A secure peer program will be developed later using Java, the code will enhance the security level by using —the DES, 3DES and AES algorithms. This program will then be run on the two clients to evaluate system performance after applying the encryption methods. At Client 1, the voice data will be encrypted using a specific encryption key and then transferred over the channel toward the Asterisk server; the server will then forward these encrypted voice data to Client 2. At Client 2, the same encryption method with the same decryption key will be used to decrypt the voice message.

1.8 Risks

Several risks are probable in this study, such as the following:

1. Difficulty in finding the appropriate research material and papers that are related to the project
2. Difficulty in finding suitable open-source VoIP clients and VoIP servers

3. Difficulty in implementing encryption/decryption that works in real time

Difficulty in integrating encryption/decryption modules into the VoIP system

1.9   Predicted Outcomes

The main expected results of this study are the following:

1. A thorough study of VoIP systems

2. A thorough study of security issues in VoIP and existing solutions

3. Implementation of VoIP service using the open-source platform

4. Implementation and integration of encryption algorithms (AES, DES and 3DES) in the open-source VoIP platform

Evaluation of the effects of encryption and decryption in VoIP applications on the quality of the VoIP service

1.10   Impact

The expected impacts of this study are the following:

1. Improved confidence of people in using VoIP services
2. Avoidance of all known and unknown types of attacks against networks
3. Reduction of the vulnerabilities that exist in the used protocols —SIP and RTP by using scanning vulnerabilities tools over the VoIP network
4. A guarantee of both the safety and security of information by detecting all VoIP attack types
5. A focus on increasing the rate of intrusion detection

1.11   Project Contribution

The security issues and aspects of VoIP services have been extensively studied and addressed by several researchers. In this project, however, a peer-to-peer secure voice calls communication system will be developed. This is considered to be an essential step in developing VoIP applications and enhancing security levels. The developed program will be the first of its type and will have clear improvements regarding peer-to-peer security. It will be applicable to any VoIP service that uses an Asterisk server, in order to make it secure by applying the DES, 3DES and AES encryption methods. It is important to mention that any data delay due to the use of these encryption methods will never exceed 1 *ms* and that this delay value can be neglected since it has not a clear effect on voice clarity and recognition.

1.12   Thesis Outlines

Chapter One is an introduction that includes the main aims, problem statement, methodology and the importance of the research, in addition to providing detailed information about VoIP systems.

Chapter Two is a literature review that describes the recent work of several researchers on VoIP systems and security issues and various aspects within these systems, including their results and brief descriptions of their methods.

Chapter Three introduces detailed introduction of the Asterisk server, including, security, set up, design, implementation and configuration.

Chapter Four describes the VoIP security mechanisms with encryption and decryption systems, along with a brief background of security within VoIP servers. Furthermore, it describes secure Asterisk server implementation based on DES, AES and 3DES.

Chapter Five explores VoIP analysis performance based on AES, DES and 3DES.

Chapter Six is the conclusion of all the work that has been introduced and achieved during the research study.

Chapter Two: Literature Review

2.1   Background

Several technologies have been developed in recent years that facilitate the communication process between individuals all over the world. Communication often takes place via VoIP, which is a common, well-known technology that transmits calls over the Internet, unlike ordinary PSTN. This technology has several advantages, such as no call limits and low costs [2].

However, an increasing need to improve and enhance VoIP technology is emerging, and this is leading to great changes within the communications field. VoIP technology depends on the available Internet protocol to divide the call into small digital packets. These packets are then transmitted via the IP [2].

VoIP technology has been extensively studied over the last few years, and researchers have agreed that security must be kept at an acceptable level to avoid vulnerabilities when data are being delivered [3]. This chapter is a comprehensive literature review of VoIP technology, especially regarding the security aspects that must be taken in consideration to achieve the best possible communication process.

2.2   Related Works

2.2.1   Security in VoIP

Significant developments have been achieved in the VoIP field, which is mainly focused on delivering calls over the Internet [6]. This technology has several advantages that make it preferable to the more common PSTN. On the other hand, this technology is faced with several challenges, including security risks, which limits the usability of VoIP for sensitive information. This issue does not seem to exist within the ordinary PSTN. Since VoIP operates via the Internet, the same security issues that exist on the Internet also exist within VoIP. In PSTN, one needs physical access to listen to voice conversations, while in VoIP, hackers can do so remotely.

VoIP is still in the beginning stages of development. However, as with any novel technology, security will be enhanced during the development for this technology. Researchers have pointed out the main security issues of VoIP applications and have given some solutions [6]. The top 10 issues of security in VoIP implementation are as follows:

1. Restricted gateway security.
2. Internet-bounded VoIP traffic.
3. Patching difficulties.
4. The security of VoIP is guaranteed only when the security of the underlying network is guaranteed.
5. Several other call processing systems operate within similar operating systems, and these systems have their specific security issues.
6. DOS attacks.
7. Snooping on calls.
8. Spam over IP telephony (SPIT).
9. Security needs are increased when the number of opened ports is increased.
10. Higher wireless security levels are needed for wireless phones [6].

Other researchers [7] have demonstrated that satisfying the security needs between both call parties—the sender and the receiver—is considered the greatest challenge in the development of VoIP systems. During the establishment of a VoIP session, several protocols are included, and these protocols should be secured against any type of attack to achieve a comprehensively secure system and a path to transmit data in a reliable way.

Researchers in [7] performed a detailed investigation of the interoperations that are performed within VoIP approach stack. They also discussed the underlying security issues in the various layers of the protocols that may result in security breaches. The stack of VoIP protocols is illustrated in Figure 5 [7].

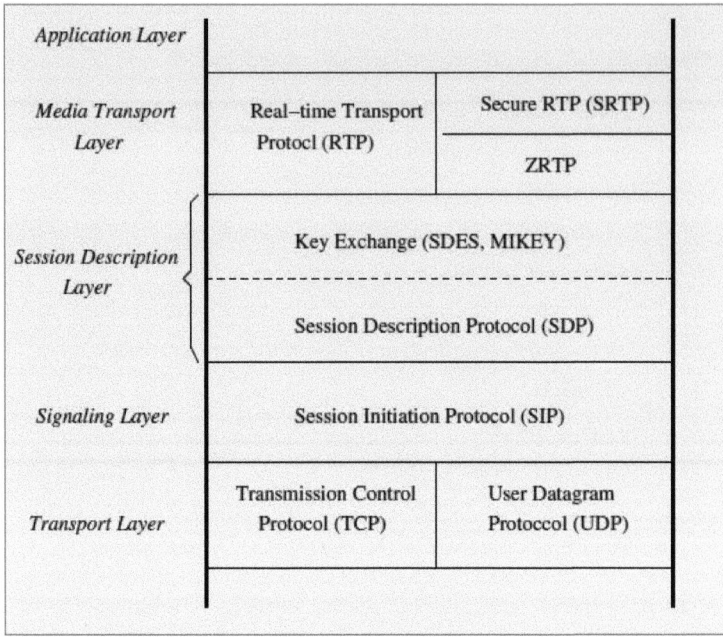

Figure 5: Stack of VoIP protocols [7].

As shown in Figure 5, the stack is separated into four primary layers: session description, signalling, media transport and transport. This separation is legal due to the fact that different protocols are present and operate within each layer. The main mission of the signalling layer is managing the process of establishing, adjusting and finishing a session between VoIP participants. Several available signalling protocols could be used for this, such as SIP, the Media Gateway Control Protocol (MGCP) and the H.232 protocols [7].

The voice data exchange process is initiated by the SDP, and the necessary cryptography is provided and achieved by using the Key Exchange Protocol. This is an essential step for creating a secure session between call participants. The datagrams of their real voices are transmitted in the transport layer, and the security within this layer is provided by the privacy and confidentiality of session keys of. It also based on the session participants' authentication. The goal of the security within the transport media is to ensure the integrity, confidentiality and authentication of the data stream [7].

According to [8], customers consider VoIP to be an attractive technology. There is an increasing rate of adoption of this new technology due to its advantages, such as minimum fees and better speeds. The vulnerabilities of the VoIP technology are similar to those that occur in the Internet environment, such as spoofing. This threat occurs when any attacker tries to illegally access a VoIP caller.

The attacker can insert a fake caller ID within the traditional VoIP to convince the call receiver that this attacker is confidence. The attacker will then be able to access the receiver's personal information, such as his account number. Several techniques can be used to achieve security within a VoIP service, including the following [8] [9]:

1. Employing a strong password
2. Encrypting the transmitted data
3. Utilising a firewall
4. Securely performing web browser configuration
5. Always keeping financial data secured
6. Updating and patching application software
7. Securely downloading software and files using authentication
8. Being careful about opening instant and email messages files

Customers are quickly moving toward adopting VoIP technologies and the Internet Multimedia Subsystem [8]. However, despite the many advantages of these technologies over PSTN, their architecture and implementation are complex. In [8], the possible risks and threats that can occur within the VoIP technology, such as DOS attacks, were investigated. They also analysed several cases to investigate the problems that may occur within a VoIP system by introducing and analysing a set of snapshots related to each problem. They concluded that the dynamic component is in included within the analysis of threat space. Several techniques and approaches can be used to verify and investigate the security protocols. Despite this, however, these techniques sometimes do not seem to work because of their high levels of complexity [8].

[10] Demonstrated that one of the most essential aspects of VoIP technology is the achievable Quality of Service (QoS). Several security measurements can be implemented to provide QoS, which results in an obvious decline of QoS. In addition, several barriers may also result within the VoIP environment, such as jitter and firewall delays and latency due to the encryption process. Time critical and low possible

interruption tolerance are two essential issues that must be taken in consideration before applying and defining the ordinary data network security measures for a VoIP environment. Modern VoIP systems depend on proprietary protocols or apply either the H.323 protocol or SIP.

Furthermore, two additional standards are available: Megaco/H.248 and MGCP. The success of the operation packet network depends mainly on several configurable parameters, such as firewall and router addresses, voice terminal, physical and IP addresses and the programmes that are used to manage and place voice calls. These parameters are dynamically recognised in case restarts the network components or when restarting VoIP calls. Therefore, there are different points between the transmitter and receiver that are vulnerable to attacks [10].

2.2.2   VoIP Security SIP-Based Networks

SIP seems to be similar to HTTP in many ways. They are both text based, they have a request-response structure and they use digest authentication. The interaction, including data exchange for several components of the network can be provided by this protocol [11]. The exchange of data in the SIP protocol is illustrated in Figure 6 [7].

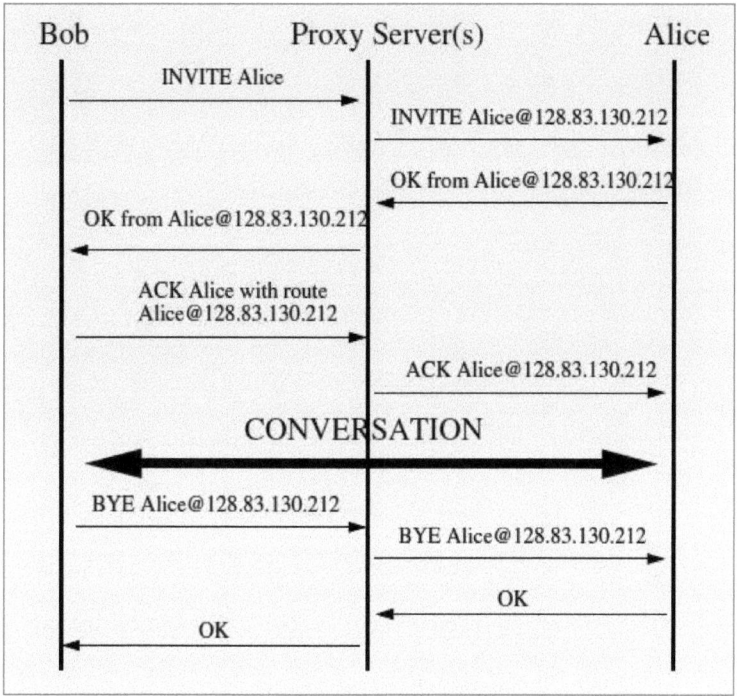

Figure 6: Data exchange of SIP protocol [7].

SIP runs over different transport layer protocols, such as the Transmission Control Protocol (TCP), UDP and the Stream Control Transmission Protocol (STCP). UDP seems to be the preferable protocol due to its advantages; for example, it provides the protection of Transport Layer Security (TLS). Alternatively, STCP is resistant to DOS, it offers mobility and multi homing support, and it is able to perform multiplexing for logical connections using only one channel. The main components of the SIP structure are summarised in Figure 7 [11].

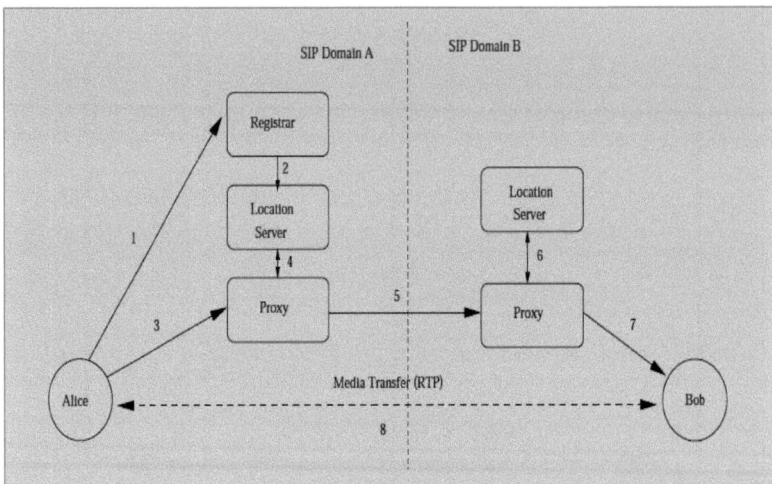

Figure 7: SIP architecture [11].

[12] Investigated security threats within VoIP systems based on the SIP protocol. They concluded that there are several common security tools, such as vulnerability assessments, protocol analyser and the utilities of security monitoring. These tools can be use to estimate probable vulnerability within several fields, such as device configuration, operating systems and network protocol applications.

Conversely, these tools also have a number of drawbacks, including their process of implementation, where it can be applied and the process that can be used to preserve and update them. Developing comprehensive tool of security for new technologies is still not easy task .In this study, they tried to determine the future issues that must be taken in considerations in order to effectively improve VoIP technology based on testing some commercial and open-source security tools [12].

The results of their investigation illustrated that the available security tools for VoIP suffer from some limitations. They also concluded that not one of these tools can deal with all probable vulnerabilities. However, some of these tools can effectively address some of these vulnerabilities. The tool chosen for use must be able to minimize all security limitations within VoIP systems and try to avoid a negative or false effect on system security [12].

In [13], it was demonstrated that SIP is considered to be a fundamental protocol that can be applied within the next generations of real-time application within communication networks. This protocol is subject to DOS attackers, and several studies have concentrated on DOS issues within systems that are based on SIP protocols [14] [15].

The results of this investigation illustrated that both flow and payload tampering attacks can be totally solved, while the remaining issue—flooding attacks—only has a partial solution. They concluded that protecting a network system from flooding attacks is the major challenge [13]. Malicious attacks can be detected by applying a change point algorithm of detection. The general structure for flooding attacks that occur within a network is illustrated in Figure 8 [13].

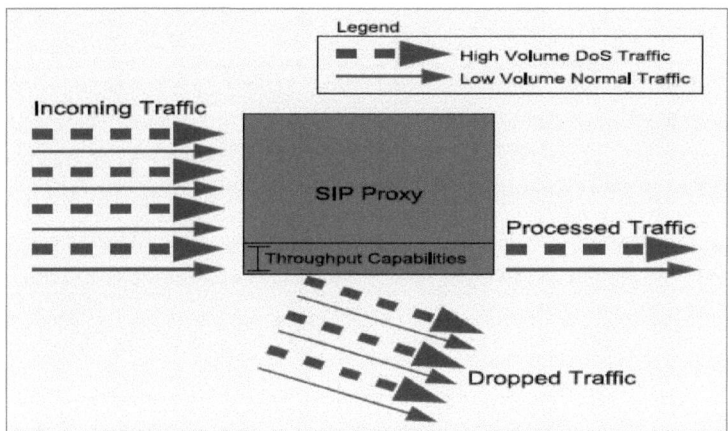

Figure 8: Structure for "DoS flooding attacks" [13].

Another study [16] found that the security mechanism deployment within a VoIP environment will result in several advantages for customers, such as confidentiality, integrity, reliability and availability. SIP is the leading signalling protocol that can be used to perform Internet calls. These researchers investigated and summarised the vulnerability problems that may occur with an SIP-based system in addition to security approaches that can be implemented and applied to solve these issues.

There must be a specific level of security that prevents attacks on information between a transmitter and a receiver, especially when this information is of great importance. The attacks on SIP in VoIP can be categorised into two main types: external and internal. There is no precise security mechanism within the SIP protocol. Two procedures are available to provide SIP security: end-end or hop-hop (Figure 9). Several available security mechanisms can also be applied by the end user, such as TLS, HTTP digest and IP security [16].

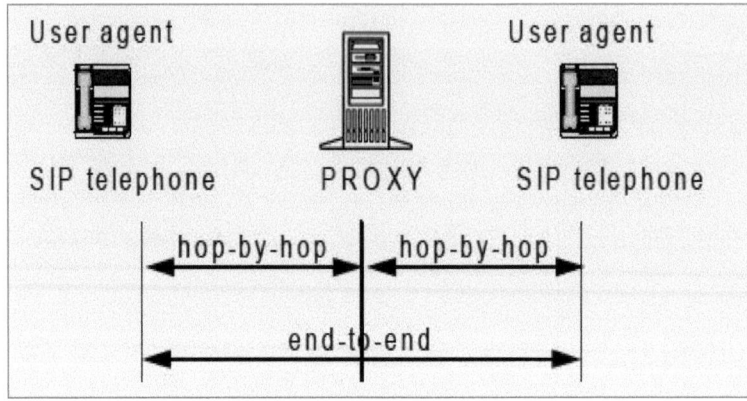

Figure 9: Mechanisim of SIP security [16].

2.2.3   VoIP Security RTP-Based Networks

In [17], researchers investigated and summarised the vulnerabilities that may occur within VoIP systems that are based on RTP. They introduced a novel method that can be use to deal with these threats effectively. Their approach was mainly concerned with managing and controlling the data stream in real time, such as the RTP protocol. The packets that contained the voice data are inserted at predictable times and can evaluated from observation. If any malicious packets are inserted within the original packets, then the test effect is maximised as much as possible. The test network that was used in this study is shown in Figure 10 [17].

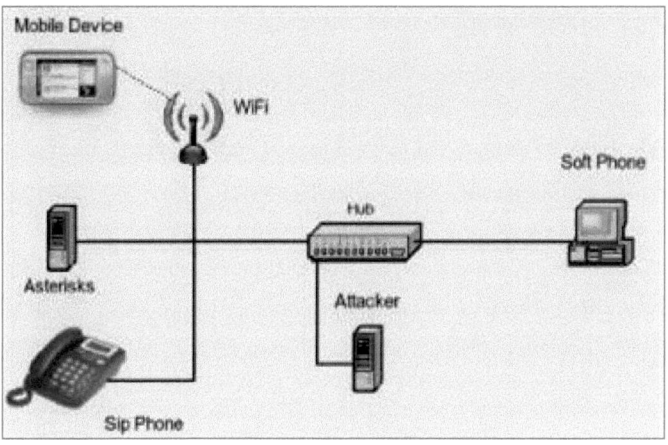

Figure 10: Experiment test network [17].

On the other hand, the inserted packets number will be minimised. In this way, the number of false or malicious packets can be minimised and the DOS detection efforts will be also avoided. They are based on sound clip modifications, so that the first packet to arrive within the stream of the RTP is stored while the later packets are discarded. In this way, an identical audio file for the listener can be generated by Wireshark. After simulating the system, the results that were obtained illustrate that these inserted packets can obstruct the RTP original stream without having any effect on system traffic [17].

Other researchers analysed the effect of applying Secure RTP (SRTP) within the VoIP system environment through two types of wireless networks: Bluetooth and 802.11-based networks [18]. To estimate this effect, they used a tool called E-Model, with SRTP security aspects, in order to specify QoS balance versus security. The results illustrated that SRTP's effect on VoIP is taken in account and never ignored.

The Mean Opinion Score (MOS) that is obtained from the E-Model for secure calls illustrates that level of QoS is unwanted for single call through wireless channels. The resulting degradation of the QoS will badly affect capacity and reduce the ability to perform calls simultaneously. These researchers also demonstrated that there are mainly three factors that affect this degradation: encryption time, authentication time and cryptography time for the received and transmitted SRTP packets [18].

2.2.4 VoIP Security IAX-Based Networks

Researchers in [19] demonstrated that media transmission and signalling are ensured within VoIP systems that are mainly based on the IAX protocol. A security services suite is also provided by this protocol, including confidentiality and authentication and Network Address Translation (NAT). Regarding the security abilities of this protocol, this protocol has been designed and created to operate within environments that are behind firewalls in addition to devices that are performing the NAT process. By using IAX, the streams between transmitter and receiver can be encrypted by applying the Dynamic Key Exchange or Rivest, Shamir, and Adelman encryption approaches [19].

In [20], it was illustrated that VoIP technology can be deployed and implemented based on centralised or distributed architecture. This is considered to be an attractive advantage or benefit to the customers when adopting VoIP technology. Regarding the centralised architecture, there are several available approaches that can be used, and IAX is classified as one of the best choices. VoIP applications are widely deployed all over the world, but there are some weakness points may limit the development and adoption for these applications. Examples of this weakness include failure, ineffective resource administration and no scalability of the system. During this study, the researchers introduced a VoIP architecture based on the IAX protocol. They also illustrated that DOS attacks can occur within VoIP systems from the Client to Server type because of failure occurrence. In a single point failure occurs at a proxy server, then all the clients within the network will be badly affect by this event [20].

2.2.5 AES, DES and 3DES in VoIP

Cryptography or the encryption is one of the familiar methods that can be used to achieve security in VoIP systems. Cryptography was originally employed to achieve security in military fields. Mathematical methods can be employed to enable both decryption and encryption processes [21].

When applying cryptography on the transmitted data; the users can reliably transfer and exchange sensitive information over insecure transmission media, such as; internet. This encrypted data will not be recognized by any DOS attacker in case it is accessed since it is encrypted and encryption key is anonymous by attacker. So; this encrypted message can be recognized by intended recipient since the decryption key is recognized at this receiver [19].

Several encryption approaches can be used in the encryption process, such as AES, DES and 3DES. These processes are used based on need. DES was developed to work best in hardware compared with software. It does plenty of "bit manipulation" in exchange and replacement boxes in each of "16 rounds". It encodes the data in "64 bit block size" and uses the key space of 56 bit to almost 72 quadrillion potential [22].

3DES is a security solution for DES, so it is more secure than DES. It is a structure that is obtained by stratifying DES three times in a sequence with three keys (K1, K2 and K3) which they have efficient length of "168 bits". Two-key 3DES is a variation of 3DES; it has an efficient key size of 112 bits but is less secure than 3DES AES; since it requires more hardware and software implementations; it is a version of the Rijndael scheme. Recently, several security aspects were used to estimate the encryption process, which are as follows [22]:

1. Whether the proposed encryption scheme is appropriate for environments that are limited in space.
2. The performance that is achieved in both hardware and software.
3. How much the scheme can resist power analysis issues and the remaining issues of implementation
4. The security of the scheme.

In [23], the security issues of VoIP systems were studies. These researchers demonstrated that there are several drawbacks that limit the use and adoption of using VoIP systems over ordinary telephone lines, including voice quality, which seems to be better via ordinary telephone connections and the lack of security in VoIP since it is performed over the Internet. The security in VoIP systems is not totally provided by VoIP technology providers; so the company that wants to adopt this technology must use it is own security mechanisms to ensure better level of security. They also demonstrated that eavesdropping may occur at any time during call transmission and reception. Therefore, security features were integrated into their study [23].

They used the RC4, AES and DES encryption process on Sipdroid to achieve secure calls with acceptable performance levels. They demonstrated that eavesdropping is probable through the communication process. Regarding to QOS, they found that the delay increases by only 0.01 ms, which does not noticeably affect the quality of the call.

31

No dropping in data rate was noticed, a loss of 8% occurred within the packets, and there is some noise due to the Sipdroid [23].

2.3   Conclusion

In summary, VoIP systems are alternatives for ordinary telephone network through which a call can be established and performed via the Internet. This system is preferable to PSTN due to it is attractive advantages, such as low costs. However, security guarantees are an essential requirement to encourage customers to use and adopting the VoIP system as a communication tool [8].

During VoIP operation, QoS requirements must be achieved to ensure an acceptable level of performance [6]. In addition, to achieve security within the VoIP environment, several issues must be taken in considerations, such as DOS attackers [6].

Several principles can be applied to achieve secure communication via the VoIP system, such as utilising a firewall and applying a strong password [8]. During the establishment of a call over the VoIP system, several protocols are included, such as SIP, IAX and RTP. The addition of security to each layer of these protocols is considered to be an essential requirement for satisfying comprehensive secured transmission and reception [7]. The encryption process could be also applied within the VoIP environment to achieve secure transmission [22, 23].

Chapter Three: Implementation of VoIP Server and VoIP Client

3.1 Introduction

The Asterisk server is the most common server for the Linux operating system. It is characterised by several advantages that make it preferable to use with a variety of applications; for example, it is reliable, free and open source. Asterisk server systems are now widely installed and used all over the world. Several features are available on the Asterisk server, such as the Private Branch Exchange (PBX), directory services, fax or telephone calls encryption, custom Interactive Voice Response (IVR) and voicemail [24].

Several hardware interfaces are provided by the Asterisk server, and these interfaces can be employed so that telephone channels can be used within the Linux box. Three main protocols are provided by the Asterisk server: SIP, IAX and H.323. Asterisk applications can be employed to connect phones or phone lines with other services or interfaces [25].

3.2 Asterisk Server Main Objectives

3.2.1   Asterisk Server Security

The Asterisk server can be defined as a channel of communication where it can carry secret information if the security aspects relayed. Various tools are used and installed to enhance the Asterisk security, such as; checker of integrity, root-kit detection, automated hardening and "Role Based Access Control (RBAC)" tool [26]. Furthermore, IP Tables can be used to make the Asterisk server secure and this can be achieved by many steps depending on the rules of IP Tables [27]. Custom chains can be also used to secure Asterisk server based on IP Tables. This began with a basic secure configuration and then it is expanded into multiple users to use the dynamic addresses of IP after that [28].

3.2.2   VOIP Asterisk Server Importance

Asterisk is an essential part of VoIP system due to the following reasons;
1. Reduced costs, as a PBX can be created via Asterisk with extremely reduced costs in comparison with traditional key PBX and key approaches [25].

2. A broad and rich base of features due to its open-source nature and its ability to be implemented using software [25].

3. Content can be dynamically deployed; several applications are provided in Asterisk throughout the AGI interface, which enables several programming platforms using Java [25].

4. Flexibility with the dial plan, which enables PBX and IVR functionality integration [25].

5. Customization due to international support, source code and configuration files [25].

Table 1 below summarizes a comparison between different types of VoIP servers.

Table 1: Comparing VOIP Servers [25, 29, 30, 31.32]

program	Operating System	protocols	Encryption	capabilities
Asterisk PBX	Solaris, Mac OS X and Linux/BSD,	IAX, SIP and H.323	SRTP and TLS	Billing, hot-desking, VoIP Gateway, conferencing, voicemail, basic accounting, and conditional logic IVR trees with
AskoziaPBX	(Linux based)	SCCP, H.323, SIP and IAX	No	ACD, Call recording, IVR ISDN, analog, Voicemail, Call forwarding Conferencing and MOH

CallMax Softswitch	Linux	H.323 and SIP	HTTPS, SSL and TLS	IP PBX, Integrated ,billing, Web Portal of Customer, Callshop module Platform and card platform of Calling

3.3 Asterisk IP PBX Server Design

The Asterisk server will be employed in designing and implementing a VoIP solution during this study, this solution can then be run on several VoIP applications. Being able to either terminate or establish new voice calls within a Local Area Network (LAN) is the major aim of this configuration. This will be achieved through the implementation of a suitable dial plan in addition to the correct equipment configuration in the laboratory of BEng. The voice call can now be customised through PBX. The Asterisk server connection with clients is illustrated in Figure 11.

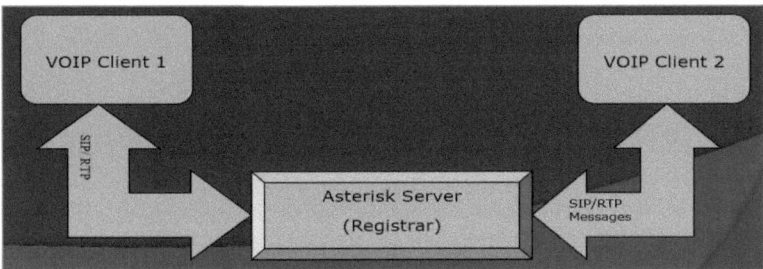

Figure 11: Asterisk server connection with clients [33].

Descriptions of the Project Requirements:

1. A Linux server that will be selected as a Asterisk server, which will be allocated to the 192.168.xx.xx IP address.

35

2. A suitable dial plan capable of 10 various extensions; 2 clients will be then involved on the windows machines.

3. A DNS server will be latterly authenticated through the machine.

4. Several VoIP calls will be made to examine the approach within LAN.

3.4   Asterisk Server Setup Implementation

The installation of Asterisk with PBX can be performed based on the following steps;

1. Go to site:www.asterisk.org/downloads

2. Download "Asterisk NOW 32-bit ISO Image"

3. Create a new virtual box and boot it up from CD-ROM that has the ISO image

4. Choose the boot ISO from the location where you saved it

5. Start the new virtual asterisk PBX server booting

When you are finished installing the packages, if you are asked to remove the CD, remove it and press the reboot button to reboot the server from the system. Your server is now completely installed and ready to us. To start configuration, use the log-in ID "root" and the password "1q2w3e".

To enter the setup mode, type: # system-config-network.

3.5   Asterisk Server Configuration
The main stages are:

- Go to the server command line and type "# service network restart".
- Configure the ethernet device to take static IP addresses.
- The server address is: 192.168.0.10

Starting Asterisk Server

To enter the graphical interface,

1. Open Internet Explorer and type: 192.168.0.10.

2. Go Inside freepbx administrator

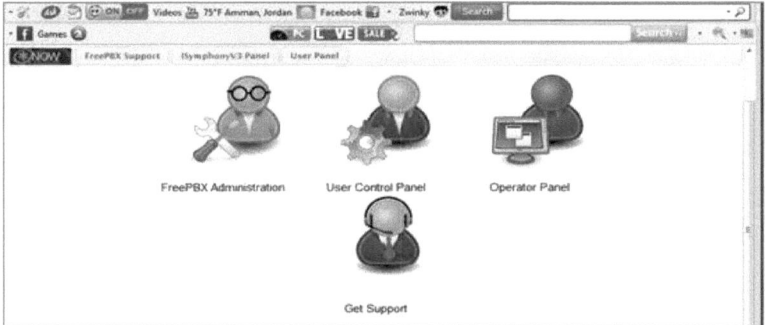

Figure 12: Enter inside freePBX administrator

3.5 Implementations SIP Extensions "Users"

Stages of the implementation of SIP Users by Using the Graphical Interface

1- Select "extensions" from applications menu

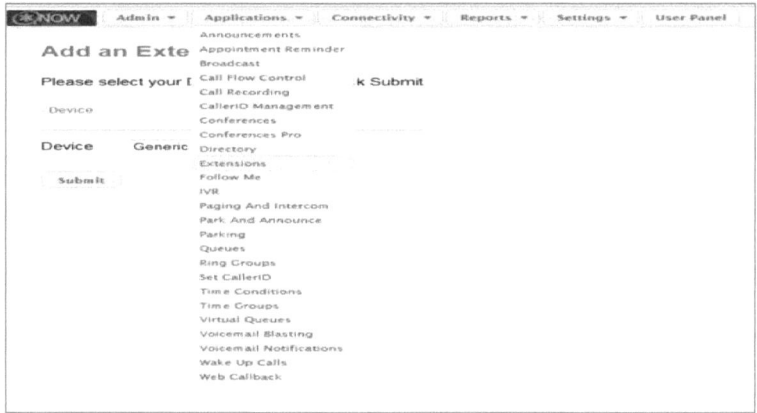

Figure 13: Implementation of sip users.

2- From the Add an Extension menu, select "generic SIP device" and then press the Submit button

3- The initial configuration is firstly entered.

The extension of SIP must be firstly get up and run. A set of information must be added, which is the extension of the user: this value can be safely selected as any value larger than 1,000.

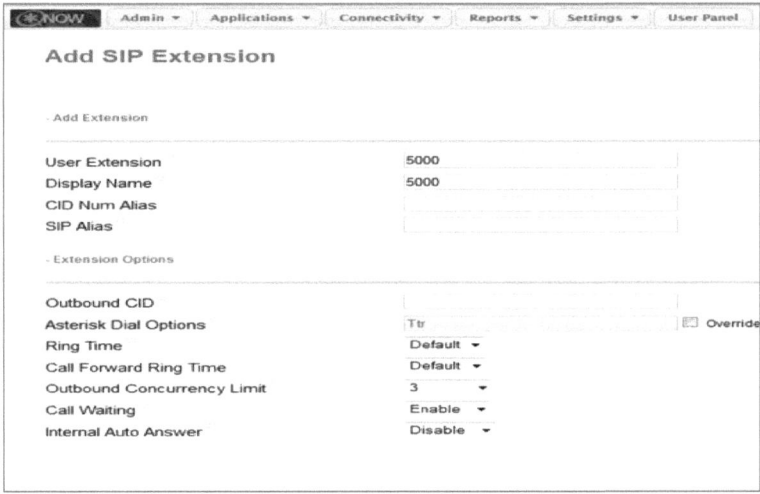

Figure 14: Enter in initial configuration

Secret: ast1234; the SIP client password must be chosen to connect with the PBX server.

4- The next step is the submission and updating; this can be performed by first selecting the submit button and then applying the needed personal configuration alterations on the PBX server. If an additional extension needs to be added, it can be done during this step. When all alterations are finished and they are ready to be applied, press the Apply Configuration Changes button.

5- A message that includes confirmation on the process of "Continue with Reload" will appear; perform this step. The novel added extension is now included within the Extension menu. The main FreePBX configuration stage is now completed and several SIP users are emerged.

3.6 SIP Java Client Code

Peers is a softphone (i.e., a software phone) and is distributed either as a binary "desktop" zip file that can be downloaded on Sourceforge on the peers project page or as a git project hosted on Github for source code. The binary zip file can be extracted on Windows, Linux or Mac, and the .bat or .sh startup script can be used to start peers. To download the peers source code, a git client must be available on your computer. Run the following:

git clone https://github.com/ymartineau/peers.git
cd peers
git checkout 0.5

Maven is a Java build tool that is pretty useful for managing dependencies amongst modules or on external libraries. You must download it, extract it and setup your environment variables if you want to build peers from a source code. Maven uses files called pom.xml to define projects and modules. The peers-demo is a command line demo project to show peers library usage with a simple example. Peers-doc is a module that contains this documentation as a docbook. It generates an html file and a pdf file. The link for the home project is http://peers.sourceforge.net/ [34]. Table 2 below summarizes a comparison between available VoIP clients.

Table 2: Comparing VOIP Clients [35, 36, 37]

program	Open Source	Operating System	protocols	Encryption	capabilities
Peers	Yes	Windows Mac ,Linux, OS, (all java supported)	SIP and RTP	May be incorporated to be SRTP or TLS	Audio calling
Sipdroid	Yes	Android	SIP	Unknown	Uses 3G, Wi-Fi, or EDGE
iCall Mobile	No	iOS v4.3+	XMPP, SIP, ICQ and AIM	TLS and ZRTP	VoIP over 3G or Wi-Fi, SMS, Messenger, XMPP Windows Live Facebook, Voicemail, AOL and Yahoo

3.7 Implementation the Sip Users on Clients Laptops

1. Open Eclipse and import the Java peers source code inside it.

2. Run the Java peers source code as a Java application to obtain the results.

3. Register the SIP client on the Asterisk server. From Edit, choose Account.

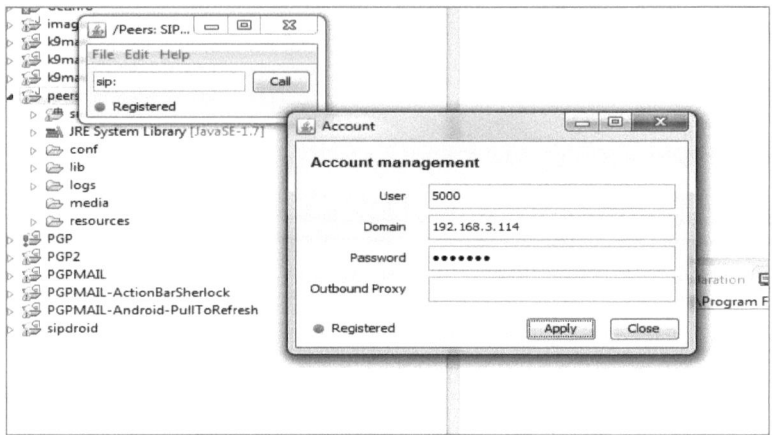

Figure 15: Asterisk server IP

Test and Monitor the Two Clients by Calling from First Java Peers to Second Java Peers (Calling From First Laptop to Second Laptop) to get inside the Asterisk server:
Asterisk –rvvv

Localhost* CLI >

If the Java peers client's registration into Asterisk server successful, the following commands will appear inside Asterisk server.

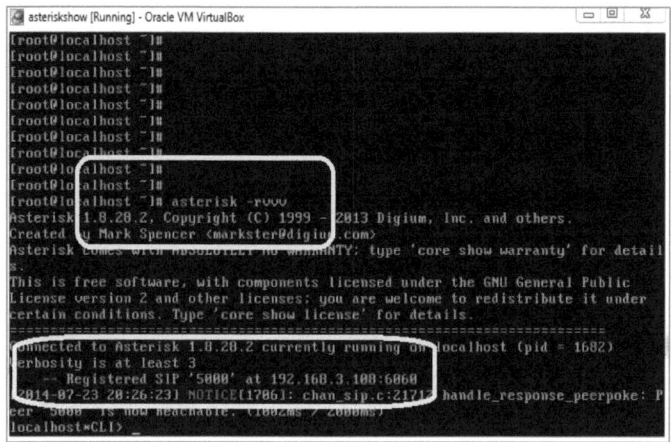

Figure 16: Java peers clients registration

Now the SIP peers will be shown using Asterisk command CLI.

Asterisk –rvvv

Localhost* CLI > show sip peers

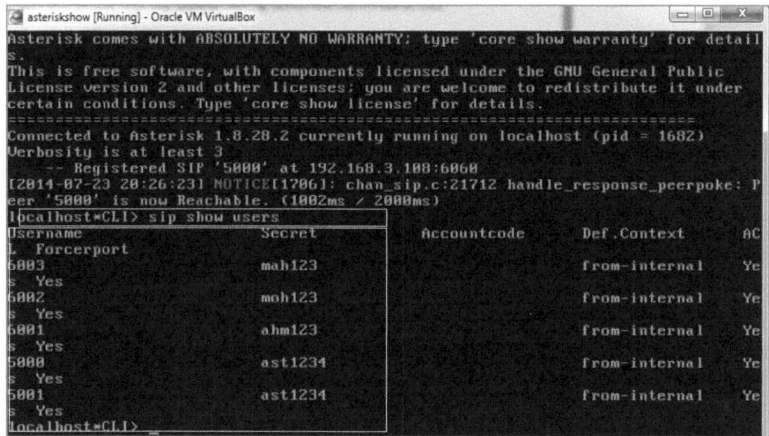

Figure 17: Sip peers

Make a call between the two SIP phones. Now, calling from one Java "peer" to anther Java "peer" is complete.

Source: Java peer: 5000 on server 192.168.3.114

41

Destination: second Java peer: 6002 on same server 192.168.3.114

From 5000 to 6002.

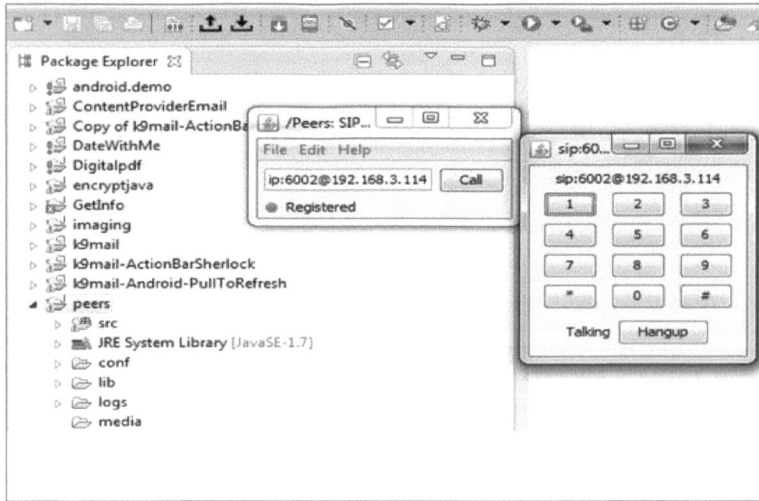

Figure 18: Calling between the two SIP phones

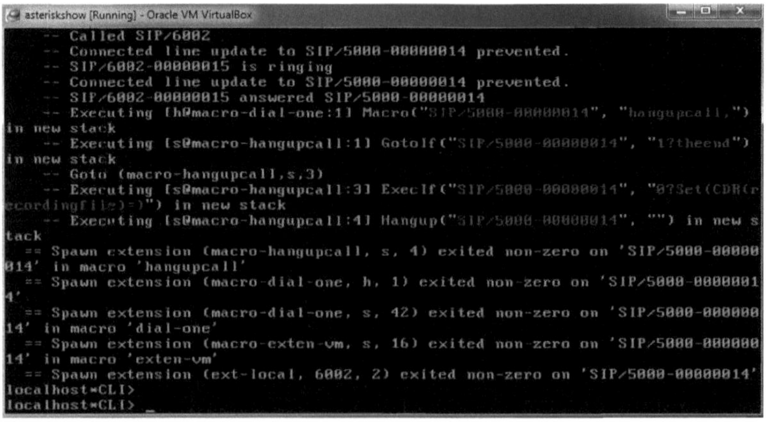

Figure 19: Calling between the two sip client

Chapter Four: Implementation of Encryption and Decryption for VoIP Communication

4.1. Introduction

After configuring Asterisk server, the peers Java client code will be adjusted to achieve secure voice communication between the two clients. The encryption/decryption will be performed using DES, 3DES and AES algorithms. This chapter introduces and investigated the encryption/decryption Java codes for these three methods.

4.2. Encrypting Voice Using DES Algorithm

DES is considered a common encryption method in which the block size is equal to 64 bits and the key that is used during execution has a 56-bit length. DES is considered a common symmetric encryption system, particularly the 16-round Feistel cipher. An identical secret key must be recognised by both the receiver and the transmitter during the communication process. This key is employed in the decryption and encryption processes, and it may also be employed in creating and confirming the message authentication code (MAC).

4.2.1   Des Encryption and Decryption Java Code

The DES Java code provides encryption and decryption. The DES modes of encryption are:

1. Electronic code book (ECB), in which the encryption of each plaintext block is performed individually.

2. Cipher block chaining (CBC), an XOR operation with a former text block of the cipher that is employed within each of the plaintext blocks.

3. Propagating cipher block-chaining (PCBC).

4. Cipher feedback mode (CFB), the encryption of the former text block of the cipher is performed here by applying the XOR operation with the block of plaintext; the equivalent text block of the cipher is now produced.

5. Output feedback mode (OFB).

4.2.2 Encrypt Function

```
final MessageDigest md = MessageDigest.getInstance("md5");
final byte[] digestOfPassword = md.digest("HG58YZ3CR9"
        .getBytes("utf-8"));
final byte[] keyBytes = Arrays.copyOf(digestOfPassword, 8);

// Generate Des Key
final SecretKey key = new SecretKeySpec(keyBytes, "DES");
final IvParameterSpec iv = new IvParameterSpec(new byte[8]);
```

Figure 20: Generating DES key

```
//cipher type :Des - CBC mod  - Nopadding
final Cipher cipher = Cipher.getInstance("DES/CBC/NoPadding");

Step 3. Initialize the Cipher for Encryption

cipher.init(Cipher.ENCRYPT_MODE, key, iv);

// final encrypted bytes
final byte[] cipherText = cipher.doFinal(plainTextBytes);

return cipherText;
}
```

Figure 21: Creating a Cipher

4.2.3 Decryption Function

```
// Des decryption function
Public static synchronized byte[] decrypt(byte[] message) throws
Exception {

        final MessageDigest md = MessageDigest.getInstance("md5");
        final byte[] digestOfPassword = md.digest("HG58YZ3CR9"
                .getBytes("utf-8"));
        final byte[] keyBytes = Arrays.copyOf(digestOfPassword, 8);

        // Generate Des Key
        final SecretKey key = new SecretKeySpec(keyBytes, "DES");
        final IvParameterSpec iv = new IvParameterSpec(new byte[8]);
        final Cipher decipher =
Cipher.getInstance("DES/CBC/NoPadding");
        decipher.init(Cipher.DECRYPT_MODE, key, iv);
        Decrypt the cipher bytes using doFinal method

        // final decrypted byte[]
        final byte[] plainText = decipher.doFinal(message);

        return plainText ;
    }
```

Figure 22: DES Decryption Function

4.2.4 Encryption and Decryption of RTP Voice Byte

DES was utilised as the cipher algorithm in this section to encrypt and decrypt the RT packet.

4.2.5 Import Des encryption and decryption into peers java client

Figure 23 Import Des encryption and decryption into peers java client

Create a new package called DES with a new DES Java class.

4.2.6 Encryption of RTP Voice Byte

Inside the "peers.rtp" package, the Java RTP session class is responsible for sending from one client to another. The Captured voice is encrypted by applying the DES encrypt function and creating an encrypted RTP packet inside the Java RTP session class

45

```
public void send(RtpPacket rtpPacket) {
        Des Des_enc = null;
        Des_enc = new Des();
            byte[] buf = null;

        if ( rtpPacket.getData().length >= 100 ){ // start encryption
        byte[] enc_rtppacket = null ;

        System.out.println("original-rtppacket          : " +
rtpPacket.getData());
        System.out.println("original-rtppacket length    : " +
rtpPacket.getData().length);


        enc_rtppacket  = Des_enc.encrypt( rtpPacket.getData());

        System.out.println("enc_rtppacket     : " + enc_rtppacket );
        System.out.println("enc_rtppacket length    : " +
enc_rtppacket.length );


        rtpPacket.setData(enc_rtppacket);
        buf = rtpParser.encode(rtpPacket);

        System.out.println("send rtppacket.data  : " + rtpPacket.getData());


        DatagramPacket datagramPacket = new DatagramPacket(buf, buf.length,
remoteAddress, remotePort);
        } // always  buf.length== 160
        }//end send function
```

Figure 24: Encryption of RTP Voice Byte

4.2.7 Decryption of Rtp voice byte

When the RTP packet arrived to the peer client, it was received by the RTP session Java class from the peers RTP package and playback was done through a computer speaker. The incoming RTP voice byte is encrypted by calling the DESede decrypt function and decrypting the RTP voice byte. Then, the decrypted voice byte is played back.

```
DES des_dec = null
des_dec = new DES();

byte[] data = datagramPacket.getData();
int offset = datagramPacket.getOffset();
int length = datagramPacket.getLength();
byte[] trimmedData = new byte[length ];

System.arraycopy(data, offset, trimmedData,0,  length );

RtpPacket rtpPacket = rtpParser.decode(trimmedData);

if ( rtpPacket.getData().length > 150 ){
            byte [] data_dect = null ;

        try {
            data_dect = des_dec.decrypt(rtpPacket.getData());
        } catch (Exception e) {
                    // TODO Auto-generated catch block
            e.printStackTrace();
        }.// DES decryption bytes
            System.out.println("Data_decrypt                 : " +
data_dect );
            System.out.println("Data_decrypt-length          : " +
data_dect.length );
            rtpPacket.setData(data_dect);

    }//end if length
```

Figure 25: Decryption of Rtp voice byte in DES

4.3.   Encrypting Voice Using TripleDES Algorithm

In a case where three of the 64-bit keys are grouped within each other, a Triple DES
(3DES) emerged with a 192-bit total length for the key. Within a private encryptor,
these 192 bits are entered at one time rather than the three grouped keys being entered
individually. This entered key will be latterly separated into three individual sub keys
using DES DLL. The keys may be padded, so a length of 64 bits results in each one of
the separated sub keys. Regarding the encryption procedure, it is identical to that used
in DES, except it must be repeated three times. The input data will be subsequently
encrypted by the first, second and third keys.

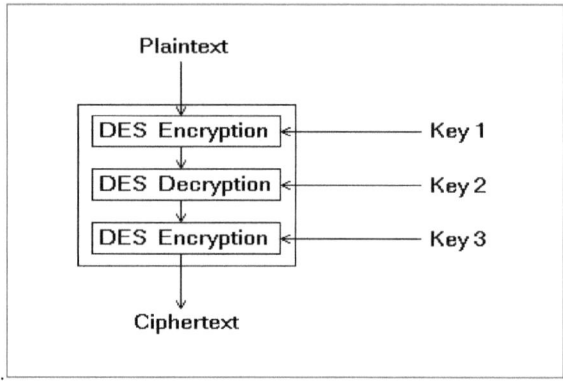

Figure 26: Data Encryption Standard

4.3.1 TripleDes Encryption and Decryption Java Code

The 3DES Java code provides encryption and decryption.

4.3.2 Encrypt Function

```
public static synchronized byte[] encrypt(final byte[] plainTextBytes)
throws Exception
{
    Step 1. Generate a DESede key using KeyGenerator

        final MessageDigest md = MessageDigest.getInstance("md5");
        final byte[] digestOfPassword = md.digest("HG58YZ3CR9"
            .getBytes("utf-8"));
        final byte[] keyBytes = Arrays.copyOf(digestOfPassword, 24);

        // Generate DESede Key
        final SecretKey key = new SecretKeySpec(keyBytes, "DESede");
        final IvParameterSpec iv = new IvParameterSpec(new byte[8]);
```

Figure 27: Genearting 3DES key using key generator.

```
//cipher type ;DESede - CBC mod  - Nopadding
        final Cipher cipher =
Cipher.getInstance("DESede/CBC/NoPadding");

        Step 3. Initialize the Cipher for Encryption

        cipher.init(Cipher.ENCRYPT_MODE, key, iv);

        // final encrypted bytes
        final byte[] cipherText = cipher.doFinal(plainTextBytes);

        return cipherText;
    }
```

Figure 28: Creating a Cipher

4.3.3 Decrypting Function

```
            // DESede decryption function
            Public static synchronized byte[] decrypt(byte[] message) throws
Exception {

                final MessageDigest md = MessageDigest.getInstance("md5");
                final byte[] digestOfPassword = md.digest("HG58YZ3CR9"
                        .getBytes("utf-8"));
                final byte[] keyBytes = Arrays.copyOf(digestOfPassword, 24);

                // Generate DESede Key
                final SecretKey key = new SecretKeySpec(keyBytes, "DESede");
                final IvParameterSpec iv = new IvParameterSpec(new byte[8]);
                final Cipher decipher =
Cipher.getInstance("DESede/CBC/NoPadding");
                decipher.init(Cipher.DECRYPT_MODE, key, iv);
                Decrypt the cipher bytes using doFinal method

                // final decrypted byte[]
                final byte[] plainText = decipher.doFinal(message);

                return plainText ;
        }
```

Figure 29: Decrypting Function

4.3.4 Encryption and Decryption of RTP Voice Byte

The cipher algorithm used in this section is 3DES to encrypt and decrypt an RTP packet.

4.3.5   Import Triple Des encryption and decryption into peers java client

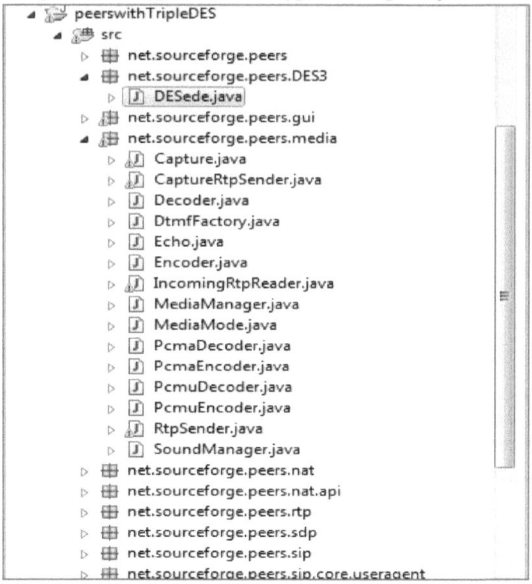

Figure 30 Import Triple DES encryption and decryption into peers java client

A new package is created called 3DES with a new DESede Java class.

4.3.6   Encryption of RTP Voice Byte

Inside the "peers.rtp" package, the Java RTP session class is responsible for sending from one client to another. The voice capture is encrypted by the 3DES encryption function and the encrypted RTP packet is created inside the Java RTP session class.

```
public void send(RtpPacket rtpPacket) {
    DESede Des3_enc = null;
    Des_enc = new DESede();
    byte[] buf = null;

    if ( rtpPacket.getData().length >= 100 ){ // start encryption
    byte[] enc_rtppacket = null ;

    System.out.println("original-rtppacket            : " +
rtpPacket.getData());
    System.out.println("original-rtppacket length     : " +
rtpPacket.getData().length);


    enc_rtppacket  = Des3_enc.encrypt( rtpPacket.getData());

    System.out.println("enc_rtppacket        : " + enc_rtppacket );
    System.out.println("enc_rtppacket length    : " +
enc_rtppacket.length );


    rtpPacket.setData(enc_rtppacket);
    buf = rtpParser.encode(rtpPacket);

    System.out.println("send rtppacket.data  : " + rtpPacket.getData());


    DatagramPacket datagramPacket = new DatagramPacket(buf, buf.length,
    remoteAddress, remotePort);
    } // always  buf.length== 160
    }//end send function
```

Figure 31: Encryption of RTP Voice Byte in 3-DES

4.3.7   Decryption of RTP Voice Byte

When the RTP packet arrived to the peer client, it was received by the RTP session Java class from the peers RTP package and was played back on a computer speaker. DESede decryption function is then used in decrypting the incoming RTP voice and the RTP voice byte was then played back as the decrypted voice.

51

```
DESede Des3_dec = null;
Des3_dec = new DESede();

byte[] data = datagramPacket.getData();
int offset = datagramPacket.getOffset();
int length = datagramPacket.getLength();
byte[] trimmedData = new byte[length ];

System.arraycopy(data, offset, trimmedData,0,  length );

RtpPacket rtpPacket = rtpParser.decode(trimmedData);

if ( rtpPacket.getData().length > 150 ){
                byte [] data_dect = null ;

        try {
                data_dect = Des3_dec.decrypt(rtpPacket.getData());
        } catch (Exception e) {
                // TODO Auto-generated catch block
                e.printStackTrace();
        } // DES decryption bytes
                System.out.println("Data_decrypt          : "  +
data_dect );
                System.out.println("Data_decrypt-length    : "  +
data_dect.length );
                rtpPacket.setData(data_dect);

        }//end if length
```

Figure 32: Decryption of RTP Voice Byte in 3-DES

4.4. Encrypting Voice Using AES Algorithm

AES is utilised by SRTP as a default cipher during both the decryption and encryption processes; this will result in a more confident flow of data.

4.4.1   AES encryption and decryption java code

Both decryption and encryption using AES functions are provided by a Java code that is related to the AES encryption algorithm. The AES algorithm is performed in the following steps:

1. AES key creation; this step includes determining the size of the key used.

2. Cipher creation.

3. The cipher is latterly initialised for encryption in order to perform the encryption process.

4. Finally, the cipher is initialised for decryption in order to perform the decryption process.

```
public static synchronized byte[] encrypt(final byte[] plainTextBytes)
throws Exception {
        final MessageDigest md = MessageDigest.getInstance("md5");
        final byte[] digestOfPassword = md.digest("HG58YZ3CR9"
                .getBytes("utf-16"));
        final byte[] keyBytes = Arrays.copyOf(digestOfPassword, 16);

        // Generate Aes Key
        final SecretKey key = new SecretKeySpec(keyBytes, "AES");
        final IvParameterSpec iv = new IvParameterSpec(new byte[16]);
        //cipher type ;Aes - CBC mod  - NoPadding
        final Cipher cipher = Cipher.getInstance("AES/CBC/NoPadding");
        cipher.init(Cipher.ENCRYPT_MODE, key, iv);

        // final encrypted bytes
        final byte[] cipherText = cipher.doFinal(plainTextBytes);

        return cipherText;
    }

   // Aes decryption function
   public static synchronized byte[] decrypt(byte[] message) throws
Exception {
        final MessageDigest md = MessageDigest.getInstance("md5");
        final byte[] digestOfPassword = md.digest("HG58YZ3CR9"
                .getBytes("utf-16"));
        final byte[] keyBytes = Arrays.copyOf(digestOfPassword, 16);

        // Generate Aes Key
        final SecretKey key = new SecretKeySpec(keyBytes, "AES");
        final IvParameterSpec iv = new IvParameterSpec(new byte[16]);
        final Cipher decipher =
Cipher.getInstance("AES/CBC/NoPadding");
        decipher.init(Cipher.DECRYPT_MODE, key, iv);

        // final decrypted byte[]
        final byte[] plainText = decipher.doFinal(message);

        return plainText ;
    }
```

Figure 33: Encryption and decryption java code in AES

4.4.2   Encryption and Decryption of RTP voice byte

AES is utilised as the cipher algorithm in this section to encrypt and decrypt the RTP
packet.

4.4.3 Import AES Encryption and Decryption into Peers Java Client

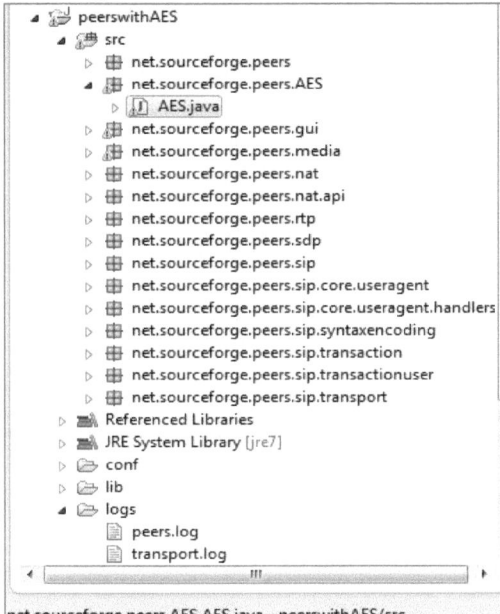

Figure 34: Import AES encryption and decryption into peers java client

A new package is created called AES with a new AES Java class.

4.4.4 Encryption of RTP Voice Byte

Inside the "peers.rtp" package, the Java RTP session class is responsible for sending the RTP packet from one client to another. The voice capture is encrypted by the AES encryption function and the encrypted RTP packet is created inside the Java RTP session class.

```
public void send(RtpPacket rtpPacket) {
    AES Aes_enc = null;
    Aes_enc = new AES();
        byte[] buf = null;

    if ( rtpPacket.getData().length >= 100 ){ // start encryption
    byte[] enc_rtppacket = null ;

    System.out.println("original-rtppacket              : " +
rtpPacket.getData());
    System.out.println("original-rtppacket length       : " +
rtpPacket.getData().length);


    enc_rtppacket  = Aes_enc.encrypt( rtpPacket.getData());

    System.out.println("enc_rtppacket     : "  + enc_rtppacket );
    System.out.println("enc_rtppacket length     : "  +
enc_rtppacket.length );


    rtpPacket.setData(enc_rtppacket);
    buf = rtpParser.encode(rtpPacket);

    System.out.println("send rtppacket.data   : "  + rtpPacket.getData());

     DatagramPacket datagramPacket = new DatagramPacket(buf, buf.length,
     remoteAddress, remotePort);
     } // always  buf.length== 160
     }//end send function
```

Figure 35: Encryption of RTP Voice Byte in AES

4.4.5 Decryption of RTP Voice Byte

When the RTP packet arrived to the peer client, it was received by the RTP session Java class from the peers RTP package and it was played back on a computer speaker. The incoming RTP voice was decrypted by the AES decryption function and it was then played back as the decrypted voice.

```
AES Aes_dec = null
  Aes_dec = new AES();

  byte[] data = datagramPacket.getData();
  int offset = datagramPacket.getOffset();
  int length = datagramPacket.getLength();
  byte[] trimmedData = new byte[length ];

System.arraycopy(data, offset, trimmedData,0,  length );

RtpPacket rtpPacket = rtpParser.decode(trimmedData);

if ( rtpPacket.getData().length > 150 ){
        byte [] data_dect = null ;

    try {
        data_dect = Aes_dec.decrypt(rtpPacket.getData());
    } catch (Exception e) {
            // TODO Auto-generated catch block
        e.printStackTrace();
    }// DES decryption bytes
        System.out.println("Data_decrypt            : "  +
data_dect );
        System.out.println("Data_decrypt-length     : "  +
data_dect.length );
        rtpPacket.setData(data_dect);

    }//end if length
```

Figure 36 : Decryption of RTP Voice Byte in AES

Chapter Five: VoIP Analysis Performance

5.1   Introduction

The designed secure VoIP system will now be evaluated for the three encryption/decryption methods. The delay will be used as performance criterion; both end to end delay and average encryption/decryption time will be estimated to ensure that encryption/decryption delay will not affect the quality experience.

5.2   VOIP Analysis Performance of RTP Encryption Based-DES Algorithm

The goal was to test the delay between an end-to-end source phone and a destination IP phone. In simple words, the extension 5000 is picked up and 6003 is dialled. After the 6003 response, the RTP voice packets will start sending and receiving the encrypted RTP packet. Both the IP phones were on the same network, and they were connected to the same Asterisk server on the same network. After some thought, it was identified that the best way to determine the time delay was to capture the timings for the first 5-10 encrypted RTP packets forwarding from the first client via the Asterisk server and the acknowledgement ACK packet from the server. Then, their time difference will be shown, and the average time delay of these encrypted RTP packets can be recorded.

To achieve the above, a tcpdump on eth0 is started on the Asterisk server. It is then run as follows: the packet inside the testing server (192.168.3.101) starts to be captured by calling from the Sip user 5000 to the Sip user 6003. The following commands are written inside server calling the following command is written as:

From the Centos server, start with: **Tcpdump -i eth0 -w des_user.pcap [38]**

The average end to end delay can be calculated using equation 1 below;

$$end\ to\ end\ delay$$
$$= \frac{\sum(received\ packet\ time - sent\ packet\ time)}{number\ of\ packets} \dots \dots (1)[39,40]$$

Then the file is opened using Wireshark.

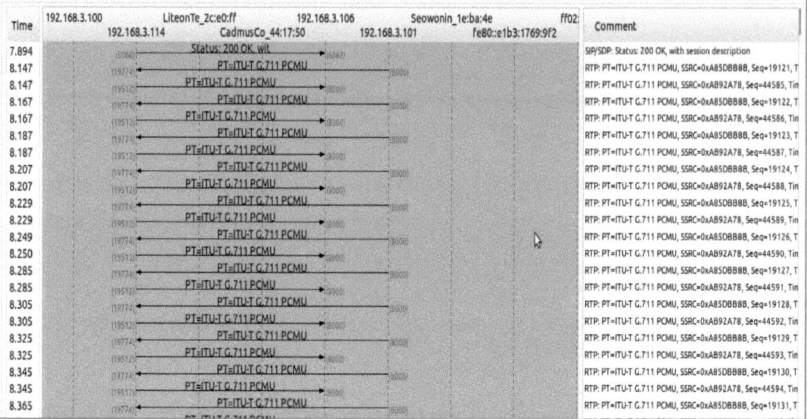

Figure 37 : Statistic graph of the rtp packet Des Algorithm

Interpreting this graph is very easy, although it might require some time if you have not done so before. For research purposes, it could be seen that:

This time: from the first client IP (192.168.3.101) to the server IP (192.168.3.114) and from the server IP to the second client IP (192.168.3.106).

5.3 VOIP Analysis Performance of RTP Encryption Based 3-DES Algorithm

The file is opened using Wireshark.

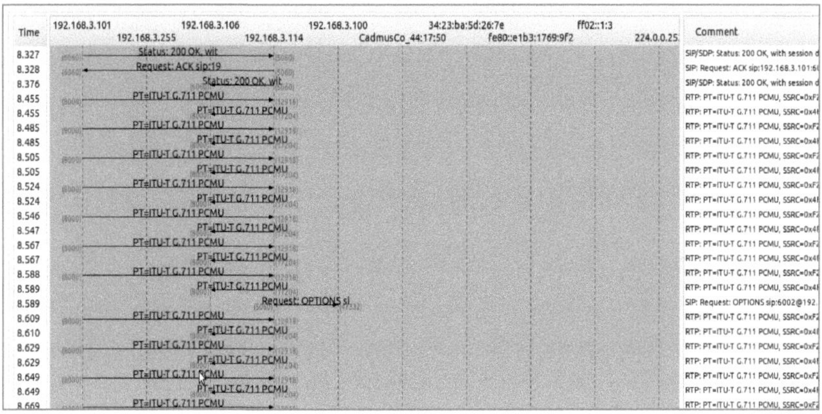

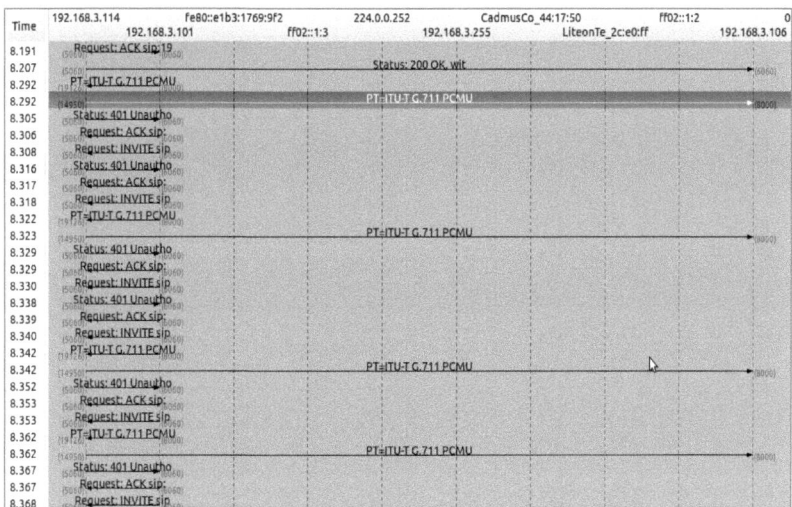

Figure 38 : Statistic graph of the RTP packet Triple Des Algorithm

Interpreting this graph is very easy, although it might require some time if you have not done so before. For research purposes, it can be seen that:

This time: from the first client IP (192.168.3.101) to server (192.168.3.114) and from server (192.168.3.114) to second client (192.168.3.106).

Figure 39 : Statistic graph of the rtp packet AES Algorithm

Interpreting this graph is very easy, although it might require some time if you have not done so before. For research purposes, it could be seen that:

This time: from first client IP (192.168.3.101) to the server IP (192.168.3.114) and from server IP to the second client IP (192.168.3.106).

As a general evaluation of the performance, the delay criterion will be used. The values shown in Figure 37, Figure 38 and Figure 39 are just an example to show how end to end delay was calculated. Table 4 below summarizes the encryption/decryption delays for AES, DES and 3DES algorithms calculated for 20000 RTP packets.

5.4   Discussion

VoIP is a transmission of the voice over an IP protocol. Since more persons are online, voice and data traffic on the Internet is increasing, resulting in the need to expand trunk networks. However, there are some constraints in VoIP applications, including bandwidth, latency, delay, reliability and security.

The average result obtained during the measurement of encryption and decryption delays shows the average delays of three schemes. Normal peers had the smallest average delay compared to peers with the encryption module. The differences in the delays that occurred were about 0:01 ms for the additions to the peers with the encryption module. Alternatively, in the peers with the encryption module, the peers with the AES and DES schemes have delays that are not much different, and peers with a 3DES scheme have the greatest end-to-end delay.

Suitable security facilities must be granted in order for real-time systems, such as IP telephony, to take off. In particular, end-to-end authentication should be possible, and this initial authentication handshake should result in session keys that can be employed in data stream protection. VoIP systems are vulnerable to all of the following attack types:

1. Application attacks by exploiting vulnerabilities in the security mechanisms for authentication.

2. DoS on media streams by introducing a large number of RTP packets or high QoS packets, as well as malformed requests.

3. Exploiting vulnerabilities in the OS or support software that allow buffer overruns and the unauthorised remote control of systems via elevated privileges.

4. VoIP systems are also open to intrusions, alterations to packet contents and destination addresses, and identity spoofing of the endpoints.

Commercial real-time systems must guard against all such attack types. Several tools, such as VOMIT, can be employed in order to enable the attacker listening to the voice streams. The results obtained from this research show that—especially from the measurement parameters of the QoS, namely delay, and from the security service parameters—the encryption schemes include AES, DES, and DES3 schemes. The testing shows that peers integrated with the encryption module are secure.

Table 3: Encryption and decryption time

		Maximum	Minimum	Average	Variance	Std. Deviation
AES	Encryption Time (ms)	5	0	0.0413	0.0476	0.2181
AES	Decryption Time (ms)	11	0	0.1930	0.1717	0.4144
DES	Encryption Time (ms)	5	0	0.0687	0.0748	0.2736
DES	Decryption Time (ms)	8	0	0.2277	0.1814	0.4260
3DES	Encryption Time (ms)	10	0	0.0803	0.0956	0.3093
3DES	Decryption Time (ms)	8	0	0.2397	0.1897	0.4355

Table 3 illustrates the maximum, minimum, average, variance and std. deviation of encryption/decryption time in DES, 3-DES and AES. The maximum decryption time=11ms for AES followed by the maximum encryption time of 10 ms for 3DES.average encryption/decryption time is always less than 1 ms and it much less as compared to the maximum value. The variance and std. deviation are also very small which means that maximum value occurs very rarely. Table 3 was used in plotting the probability for both decryption and encryption time using AES, DES and 3-DES methods. MATLAB software program was used to perform this part of the project. The outcome results will be introduced and discussed now figure 40 and figure 41 below illustrate the probability of encryption and decryption time in DES algorithm.

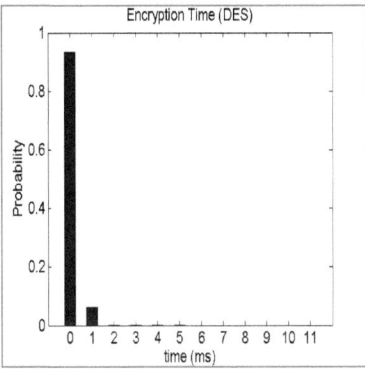

Figure 40 : probability of encryption time in DES.

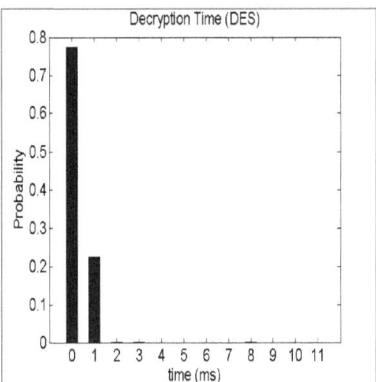

Figure 41 : probability of Decryption time in DES.

As shown in Figure 40 and Figure 41; the encryption time is less than 1ms for more than 90% of the time and it is equals to 1ms for very little time. The decryption time is also less than 1ms for 78% of the time and it is equals 1ms for few times.

The following two figures are outcome for 3DES algorithm.

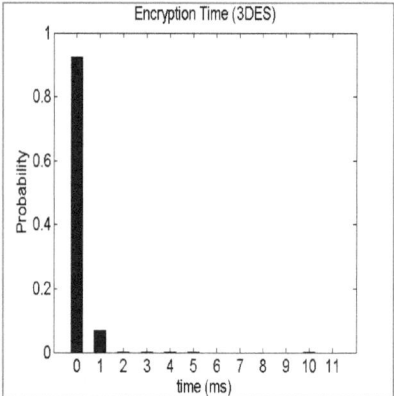

Figure 42 : probability of encryption time in 3DES

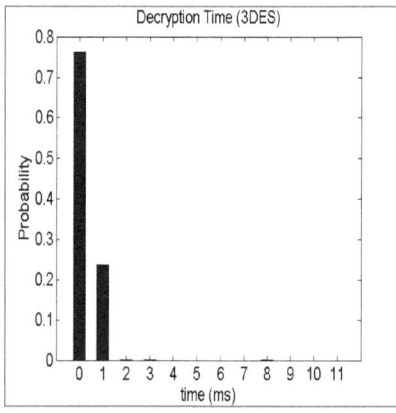

Figure 43 : probability of Decryption time in 3DES

As shown in Figure 42 and Figure 43; the encryption time is less than 1ms for 90% of the time and it is equals 1ms for very few times. The decryption time is less than 1ms for 80% of the time and it equals 1ms for 20% of the time.

AES algorithm results are shown below.

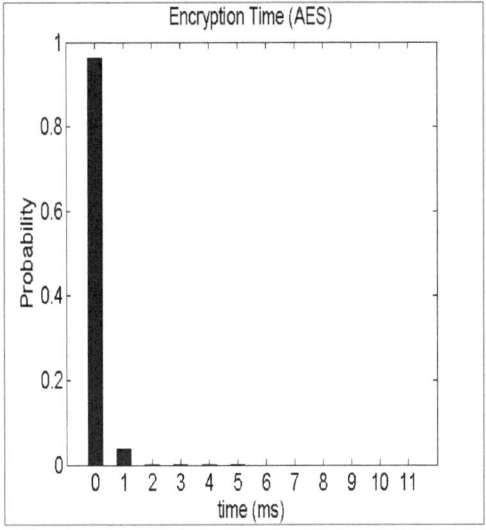

Figure 44: probability of ecryption time in AES.

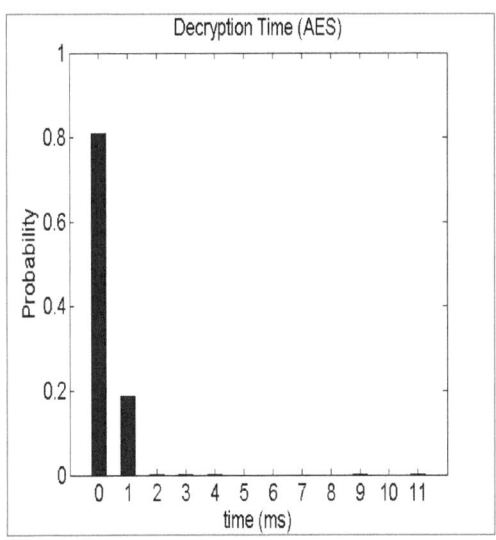

Figure 45 : probability of Decryption time AES

As shown in Figure 44 and Figure 45; the encryption time is less than 1ms for 98% of the time and it is equal 1ms for very few times. The decryption time is also less than

1ms for 80% of the time and it equal 1ms for 20% of the time. It can be concluded that for DES, 3DES and AES; the encryption and decryption time is less than 1ms for most of the time. This will result in adding a maximum of 2ms delay as a result of applying encryption/decryption. This value can be neglected. So the quality of experience does not affected by encryption/decryption.

Table 4 illustrates the end-to-end delay for original peers and AES, DES and 3DES encryption methods. The end-to-end delay includes: voice capturing, encoding, and encryption, transmission, receiving, decoding and playing back. The end-to-end delay for the original peers is equal to 18ms, while it is 20ms for AES, 19ms for DES and 20ms for 3DES. This means applying encryption/decryption only increases the end-to-end delay by a maximum of 2ms.

Table 4: Average end to end delay

	Original peers	AES	DES	3DES
Average	18ms	20ms	19ms	20ms

Chapter Six: Conclusion and Future Works

6.1 Conclusion

The usage of applications that allow real-time voice/video streaming is growing exponentially. Real-time collaboration mechanisms are likely to be integrated into the predominant applications of today, such as Microsoft Office and development tools. The need for security in real-time streams is critical to the success and continued usage of these applications.

While many commercial VoIP offerings do not provide adequate security, at the protocol specification level, various security facilities exist for the protocols of VoIP technology, including RTP and SIP. Encryption is nontrivial, and care must be taken to ensure that security is not compromised through poor designs or flaws in implementation. It is therefore recommended that the systems use well-known, publicly analysed, cryptographic techniques. This study shows that IPsec protocol is widely used in VoIP networks. AES, DES and 3DES in CBC mode for encryption algorithms are used in this protocol. In our study, the performance results and the encryption/decryption processing delays for AES, DES and 3DES in CBC mode were introduced.

The encryption module generated from AES, DES and 3DES encryptions can be integrated well with peers by encrypting the RTP packet to be transmitted on a VoIP network. The measurement performance of a security service for peers integrated with an encryption module can be done well, indicating that the integrated encryption module works well. The QoS result from this measurement enlarged the delay by more than 0.01 ms for peers integrated with an encryption module; meanwhile, for packet loss and throughput, there are no significance changes.

Peers with an encryption module can overcome a passive attack, because communication data captured by Wireshark cannot be decoded well; this happens due to the process of encrypting the data. Standard RTP provides no support for integrity protection or source origin authentication. The RTP specification provides support for the encryption of both RTP and RTCP packets. All octets of the RTP data packets, including the header and the payload, are encrypted. The default encryption algorithm for RTP is DES in CBC mode. Advances in processing capacity have rendered DES weak, so an effective encryption method, such as AES or DES, is recommended to be selected during the implementation phase. AES with a 128-bit key is optimal for real-time systems.

6.2 Future Works

According to the results, it can be suggested that when developing VoIP protocol, and in developing new and stronger encryption algorithms for IPsec, these low-cost and new generation VoIP applications can be used in order to provide efficient encryption/decryption processing delays (less than 1 ms). Furthermore, the provided Java program can be applied to any VoIP application that uses the Asterisk server in order to enhance the peer-to-peer security level within this application. Another point that can be considered as a future work is to develop an Android application that is able to be applied on Sipdroid clients.

References:

[1] S. Phithakkitnukoon, R. Dantu, and E. Baatarjav, "VoIP security attacks and solutions," Information Security Journal: A Global Perspective, vol. 17, no. 3, pp. 114-123, 2008.

[2] J. Dudman, "Voice over IP: what it is, why people want it, and where it is going," JISC Technology and Standards Watch, pp. 70-83, September 2006.

[3] H. Ingo, "Session initiation protocol (SIP) and other voice over IP (VoIP) protocols and applications," pp. 1-20, January 2007.

[4] A. D. Keromytis, "A Comprehensive Survey of Voice over IP Security", IEEE Communications Surveys & Tutorials, pp. 514 – 537, vol. 14, no. 2, 2011.

[5] T. Zourzouvillys and E. Rescorla, "An introduction to standards-based VoIP: SIP, RTP, and friends," IEEE Computer Society, vol. 14, no. 2, pp. 69-73, April 2010.

[6] M. Ruck, "Top Ten Security Issues Voice over IP (VoIP)," Design data, technology consultants and network engineer, 2010.

[7] P. Gupta and V. Shmatikov, "Security analysis of voice-over-IP protocols," IEEE Computer Security Foundations Symposium, pp. 49 - 63, July 2007.

[8] M. Desantis, "Understanding voice over internet protocol (VoIP)," US-CERT, a Government Organization, pp. 1-5, 2008.

[9] R. Baumann, S. Cavin and S. Schmid, "Voice Over IP - Security and SPIT", Swiss Army, FU Br, vol. 41, pp. 1-34, 2006.

[10] R. Kuhn, T.H. Walsh, and S. Fries, "Security considerations for voice over IP systems," SP 800-58, US National Institute of Standards and Technology, December 2003.

[11] A. Keromytis, "Voice over IP: risks, threats and vulnerabilities," Cyber Infrastructure Protection, May 2009.

[12] S. Mcgann and D. Sicker, "An analysis of security threats and tools in SIP-based VoIP systems," Second VoIP security workshop, pp. 1-8, April 2005.

[13] S. Ehlert, D. Geneiatakis, and T. Magedanz, "Survey of network security systems to counter SIP-based denial-of-service attacks," Science Direct, vol. 29, pp. 225-243, March 2009.

[14] F. Huici, S. Niccolini and N. Heureuse, "Protecting SIP against Very Large Flooding DoS Attacks", Global Telecommunications Conference (GLOBECOM 2009), pp. 1-6, November 2009.

[15] G. Ormazabal, S. Nagpal, E.Yardeni and H. Schulzrinne, "Secure SIP: A Scalable Prevention Mechanism for DoS Attacks on SIP Based VoIP Systems", IPTComm, pp. 107–132, 2008.

[16] D. Geneiatakis, G. Kambourakis, T. Dagiuklas, C. Lambrinoudakis, and S. Gritzalis, "SIP security mechanisms: A state-of-the-art review," Proceedings of the Fifth International Network Conference (INC 2005), pp. 147-155, February 2005.

[17] M. Adams and M. Kwon, "Vulnerabilities of the real-time transport (RTP) protocol for voice over IP (VoIP) traffic," In 6th IEEE Conference on Consumer Communications and Networking Conference (CCNC 2009), pp. 1-5 , October 2008.

[18] A. Passito, E. Mota, and E. Mota, "Analysis of secure RTP protocol on voice over wireless networks using extended MedQoS," Proceedings of the 2009 ACM symposium on Applied Computing, pp. 86-87, 2009.

[19] A. Lazzez, "VoIP technology: security issues analysis,"
International Journal of Emerging Trends & Technology in Computer Science (IJETTCS), vol. 2, December. 2013.

[20] A. Lazzez, Q. Fredj, and T. Slimani, "IAX-Based Peer-to-Peer VoIP Architecture," arXiv preprint arXiv:1310.5805, May 2013.

[21] H. Ayushi, "A symmetric key cryptographic algorithm," International Journal of Computer Applications, vol. 1, no. 15, pp. 1-4, August 2010.

[22] H. Alanazi, B. Zaidan, A. Zaidan, H. Jalab, M. Shabbir, and Y. Al-Nabhani, "New comparative study between DES, 3DES and AES within nine factors," Journal of Computing, vol. 2, no. 3, pp. 152-157, March 2010.

[23] A. Wahab, R. Bahaweres, M. Muhaemin, and R. Sarno, "Performance analysis of VoIP client with integrated encryption module," In 1st IEEE Conference on Communications, Signal Processing, and their Applications (ICCSPA 2013), pp. 1-6, February 2012.

[24] P. Mahler, "VoIP Telephony with Asterisk," Signate, 2005.

[25] M. Spencer, "Introduction to the Asterisk open source," Linux Support Services Inc., 2002.

[26] D.Gomillion and B. Dempster, "Building Telephony Systems with Asterisk", Packet Publishing, pp. 1-157, 2005.

[27] L. Song, "Securing Your Asterisk VoIP Server with IPTables", 2014, available at: [Online]: http://blog.ls20.com/securing-your-asterisk-voip-server-with-iptables/

[28] DarthJDG, "Iptables setup to secure Asterisk server", 2013, available at: [Online]: http://codebin.co.uk/blog/iptables-setup-to-secure-asterisk-server/

[29] Digium, "Asterisk Quick Start Guide ", *the Asterisk Company*, April. 2012.

[30] Dialogic. "PBX Integration for Asterisk-Based IP Media Servers Using Dialogic Media Gateways", *Dialogic making innovation thrive*, Sep. 2009.

[31] Speedflow, "CallMax class5 soft switch", *Speedow Communications Ltd*, June.2014.

[32] Naveen, "AskoziaPBX 2.2 Hardware 1.1 Bundle Offer(8 Ip Phones with IP PBX)", April. 2014.

[33] S. Nawaz, M. Niebur, S. Schuff, and A. Siddiqui (2009, May 16). Embedded VoIP phone [Online]. Available: http://www1.cs.columbia.edu/~sedwards/classes/2009 /4840/reports/VOIP-presentation.pdf

[34] Peers. (2012). Peers Java SIP softphone [Online]. Available: http://peers.sourceforge.net/user_manual/

[35] Alternativeto. (July. 2014). *Icall* [online]. Available at: http://alternativeto.net/software/icall/

[36]N. Unuth. (2014). What's VoIP Clients [online]. Available at: http://voip.about.com/od/glossary/a/VoipClient.htm

[37] Nomado. (2012). Softphone Sipdroid configuration Guide [online]. Available at: http://www.nomado.eu/eshop/product_info.php/cPath/40/products_id/193

[38] Y. Tsang, M. Yildiz, P. Barford and R. Nowak, "On the Performance of Round Trip Time Network Tomography", IEEE International Conference on Communication, vol. 2, pp. 483-488, June 2006.

[39] J. L. A. Fonseca and M. A. Stanton, "A Methodology for Performance Analysis of Real-Time Continuous Media Applications", 12th International Workshop on Distributed Systems, October 2001.

[40] M. M. Alani, "Mathematical Approximation of Delay in Voice over IP", International Journal of Computer and Information Technology, vol. 3, January 2014.